HYBRID CARS

The motive power behind the world's salvation?

by

Owen Jones

Owen Jones

Copyright

Hello and thank you for buying this book called 'Hybrid Cars - The motive power behind the world's salvation?'

I hope that you will find the information helpful, useful and profitable.

Hybrid cars, and indeed, all electric vehicles are playing a rôle in the salvation of the planet's eco-system. At least, in respect to the world supporting human life. However, hybrid cars will one day be replaced by vehicles that use no irreplaceable energy at at all, once the technology is there. It is likely that there will be at least two such power sources in the near future: electricity and hydrogen – but who knows what else will be available?

Hybrid and electric cars are revolutionizing the automotive landscape, offering a myriad of advantages that extend beyond the traditional internal combustion engine. One of the primary benefits is environmental sustainability. Electric cars produce zero tailpipe emissions, significantly reducing air pollution and carbon footprint. As the world grapples with climate change, the shift towards electric vehicles is a crucial step in mitigating environmental impact.

Furthermore, these vehicles contribute to energy efficiency. Hybrid cars combine an internal combustion engine with an electric motor, optimizing fuel consumption and reducing reliance on fossil fuels. Electric cars, relying solely on electric power, boast high energy efficiency and cost savings over time. With advancements in battery technology, electric cars now offer extended ranges and faster charging times, addressing concerns about range anxiety.

Beyond environmental and economic advantages, hybrid and electric cars provide a smoother and quieter driving experience. The instant torque delivery in electric motors enhances acceleration, making these vehicles not only eco-friendly but also performance-oriented. As governments worldwide incentivise the adoption of cleaner technologies, the advantages of hybrid and electric cars position them as the driving force in the future of sustainable and efficient transportation.

The information in this ebook on various aspects of hybrid vehicles and related subjects is organised into 16 chapters of about 500-600 words each.

I hope that it will interest those who are interested in hybrid vehicles.

As an added bonus, I am granting you permission to use the content on your own website or in your own blogs and newsletter, although it is better if you rewrite them in your own words first.

If you have any feedback, please leave it with the company you bought this book from.

Thanks again for purchasing this book,

Regards,

Owen Jones

Table of Contents

Short History of Hybrid Cars 7
How Do Hybrid Cars Work? 11
Hybrid Cars in the Early
Twenty-First Century 15
Hybrid Cars and City Living 19
Hybrid Cars - The Vehicles of
the Future 23
Hybrid Cars and Hybrid Trucks 27
Hydrogen Fuel Cars - Do They Exist? 31
New Hybrid Vehicles 35
Hybrid Electric Engines 39
The Hybrid Car and Petrol Prices 43
Electric Hybrid Cars 47
Run Your Car on Water 51
Hybrid Car Myths 55
Hybrid Cars and the Energy Crisis 59
Hybrid Cars vs Plug-in Hybrid Cars 63
The Future Landscape of Hybrid Cars 67
Contact Details 73

A Short History of Hybrid Cars

The first problem with working out which was the first hybrid vehicle, is deciding what the term 'hybrid vehicle' means. For instance, a barge being pulled by a horse with the current of the canal might be thought of as a hybrid vehicle. However, most people these days will agree that a real hybrid vehicle utilises a 'rechargeable energy storage system' or an RESS and some other type of liquid or solar energy source.

For instance, this could define a vehicle that uses one form of propulsion, such as an internal combustion engine as its foremost form of propulsion, whilst that engine recharges batteries that can also be used to power an electric engine by the use of a combination of solar and dynamo power generation.

I am certain that it will surprise ninety percent plus of people to hear that the history of hybrid vehicles is almost as long as the history of the automobile itself. Porsche is a famous manufacturer of costly sports cars, but in 1898 Ferdinand Porsche, a young Czech student, designed the Loher-Porsche one-cylinder internal combustion engine.

However, this engine was used to drive an electric generator, the electricity from which was employed to power electric motors which were affixed to every of the four wheels. The petrol engine was used only to generate electricity for the electric motors in this early example.

This early hybrid was presented at the World Exhibition in Paris in 1900 and was capable of travelling at 35 mph (56 kph). In 1901, Porsche drove it himself to win the Exelberg Rally. After this they sold more than 300 units of their primitive hybrid car. Mass production had not been invented yet and rich people were still sceptical about the new, malodorous technology.

1959 was the next landmark in the history or hybrids because petrol was cheap and few people, if any, foresaw the future for the world and the environment. Anyway, the car invented, the Henney Kilowatt used the early transistors or those days to control the flow of electricity. This was the true precursor to modern hybrid cars.

One of the inventors of the Henney Kilowatt was Victor Wauk and he was involved in the process of experimenting with electric cars in the Sixties and Seventies. Occasionally, he is called the Godfather of Hybrid Vehicles.

It is fairly remarkable, but the regenerative braking system used by contemporary hybrids to help recharge a hybrid's batteries was invented in 1978 by the electrical engineer, David Arthurs..

It then took until president Bill Clinton took the initiative to instigate the Partnership for a New Generation of Vehicles in 1993. It involved the Dept of Energy, Chrysler, ford, GM and one or two others. G. W. Bush replaced this scheme with his own Freedom Car Initiative in 2001.

This initiative was designed to fund extraordinarily risky or problematic projects for the development of hybrid cars. It has taken us over 100 years to revive the initial hybrid concept but we only did that because we were compelled to do it.

Hybrid Cars

How Do Hybrid Cars Work?

Basically, hybrid electric cars have two engines: a conventional petrol or diesel engine (the same as you would see in any modern car and an electric, battery powered engine, as you may find in a milk float or a forklift truck. The magical difference is that the car's on board computer judges which engine is necessary to provide the power needed by the driver and turns it on.

Consequently, if you are accelerating to cruising speed for motorway driving; going up hill or overtaking, the car will probably use its liquid fuel engine and then as you slacken off the accelerator to, say,
cruise down the motorway; go down the other side of the hill or to drive in heavy traffic, the computer will turn off the liquid fuel engine and turn on the electric engine.

The electric engine can be regarded as free to run, because it runs off batteries which are recharged by the car while it is using petrol or diesel and at some other times, such as while it is braking (and the alternators are recharging in both modes). You should never need to recharge your car's batteries overnight as they do with forklift trucks.

There are basically two types of hybrid cars: the semi hybrids and the full hybrids.

The semi hybrids have the same type of set up: two engines, one running on liquid fuel and the other running on batteries, but the electric motor is not capable of running the car on its own. It is there to 'assist' the petrol or diesel engine.

In this type of hybrid, the electric motor is known as an 'assist'. These semi hybrids will save money on fuel, but when the car is moving, you are burning fuel all the time.

The main difference when it comes to the full hybrid is that both engines are capable of powering the car independently. When you are running on electricity, you are running at zero expense to your wallet and at zero cost to the environment, unless you are really pushing the car and then both engines may be working in union.

This changing of power sources is done robotically

without any interference from the driver. In the case of the Prius, for instance, this extraordinary achievement is accomplished by what Ford calls its Hybrid Synergy Drive. Other companies have their equivalent to the HSD.

In order to gain the most out of these full hybrids, you actually need to be doing an 'average amount' of driving under 'average' or 'mixed' circumstances. For example, if you are driving in traffic, the car will want to use the electric engine, but if all you are doing is advancing slowly in inner city traffic jams, the batteries will soon become exhausted and so you will be driving on liquid fuel all the time. Ths rather negates the foremost reason for spending a great deal extra on a hybrid in the first instance.

The car has to travel on open motorways in order to recharge its batteries so that it can utilise them when it gets back into town. If you only drive in town traffic, you may be better off getting a little run around instead.

Hybrid Cars

Hybrid Cars in the Early Twenty-First Century

You would probably be surprised to know that hybrid cars have been with us since the very earliest days of automotive history, yet you will not be surprised to be informed that the technology has come on in leaps and bounds over the last ten years. In fact, the technology in hybrid cars has reached the level where fuel consumption in a hybrid can be half that or even less than in a conventional internal combustion engine only car.

Half or even just less than half of people, under some circumstances, do most of their driving in town, especially in busy traffic. In fact, while a hybrid is either cruising on electric or stopping and starting in thick traffic, there will be no consumption of petrol or diesel whatsoever, which means that you are driving 'without cost', until the batteries run out.

I put without cost in parentheses because a hybrid

vehicle is still considerably more expensive than a car with a traditional engine. However, even given the added price of a hybrid vehicle, you can save that excess price during the average life time of the vehicle because of the saving on fuel, but you must pay up front, which is like giving a free loan to the vehicle manufacturer. Of course, the higher the price of fuel, the sooner you will recoup your extra initial outlay on the price of the hybrid, and since oil is cheap now, the extra cost is difficult to justify.

Perhaps you are thinking that hybrid vehicle manufacturers are profiteering on the backs of car drivers who want to do their bit for the environment. Well, you would not be alone in thinking that, but the cost of progressing hybrid technology was and still is expensive and someone has to pay for it.

That means you and me, the end-users. Well, that is standard, yet in some countries, the government has stepped in to subsidise individuals who buy a hybrid vehicle, because governments all around the world are under an oath to reduce their country's carbon footprint.

That means that now is as good a time as any to buy a hybrid vehicle.

There are other reasons why a hybrid is costly though. The hybrid actually has two engines. A hybrid has a traditional engine run on traditional fuels, yet it also has an electric engine which runs off costly batteries. It is not that you have to replace the batteries as with a radio. The batteries are expensive because they are very heavy duty rechargeable units.

The technology for recharging these batteries is also ground-breaking. The car uses traditional alternators to recharge them but it also employs braking power to recharge them too. Not just that but the car's on board computer switches between the two motors automatically, depending on the quantity of power that is required by the driver.

Keep an eye on the hybrid car market because prices are declining and coupled with subsidies, the time to buy a hybrid car might come sooner than you think.

Hybrid Cars

Hybrid Cars and City Living

There is a variety of reasons why you might want a hybrid car. You may like a hybrid car in order to slash your ever increasing petrol bill; in order to reduce your personal effect on the environment or you may merely want the kudos of driving a car that is at the vanguard of technology. Obviously, it could be for a combination of all three reasons as well.

Hybrid cars have been about for around ten years and so the technology is fairly well advanced. The thing to remember is that hybrids are not performance cars in the conventional sense of the word. In the perspective of cars, the word 'performancc' usually refers to 'high speed', but hybrid cars are performance cars because they save more than eight percent on the fuel bill.

They create this saving by basically using two engines. The one engine is a conventional internal combustion engine (ICE) and the other is an electric motor. Both engines deliver their power through the same mechanical means to the wheels. The ICE generates electricity and supplies it to a battery, like any car does, however, a hybrid car can use this battery power to drive the car as well.

The electricity is generated by alternators and the braking system. Regenerative braking supplies a substantial amount of power to the batteries. In fact, so much so that under regular driving conditions, the batteries do not have to be charged from the national grid.

The 'early' hybrids used the electric motor just as an 'assist'. In other words, whilst the petrol engine would normally require to get revved up to produce enough power to overtake or go up hill, the electric motor would kick in to assist it, thereby saving fuel, but the petrol engine is in fact running all the time. This a kind of halfway hybrid. The Honda Insight was one of these.

However, a full hybrid will use one or the other or both of the engines, depending on how its computer best interprets the power needs of the driving conditions. The driver has no decisions to take, engines are turned on and off automatically and seamlessly by the car's on board computer. Examples of this sort of full hybrid are the Toyota

Prius and the Ford Escape Hybrid.

Although most individuals think of hybrids as new technology, the first hybrid car was manufactured more than a hundred years ago. Contemporary hybrids are around ten years old and the technology is improving quickly. However, what really has to happen now for hybrids to make a real effect on the amount of oil that the West uses is for the prices to come down.

And I mean actually come down a great deal. Hybrid cars are way too expensive for the average driver. If manufacturers reduced the price of the cars, more people would buy one which would stimulate the economy and aid the balance of payments deficit to say nothing of the effect of burning less fossil fuel would have on the environment.

Hybrid Cars

Hybrid Cars - The Vehicles of the Future

A new kind of hybrid has entered the English language. Not so long ago, if you heard a snippet of conversation containing the word 'hybrid' you would have assumed that it either referred to a rose, as in 'F1 Hybrid' or possibly a wolf-dog variety. However, nowadays, someone saying the word 'hybrid' is more likely to be referring to a vehicle.

A new sort of car that has two engines and burns far less petrol or diesel than its antecedents because it relies on the contemporary technology of solar or / and dynamo energy. However, the idea of a hybrid car is not new at all. One of the first cars, well over a hundred years ago was a hybrid. In fact, that early hybrid car also used petrol and electricity from batteries.

Modern hybrid vehicles also create use of oil derivatives and electricity stored in batteries as sources of power. In essence, a hybrid car will use its petrol engine when the driver requires power, for instance whilst overtaking or going up hill, but it will automatically switch to the electric motor when the car is at a cruising speed or creeping through inner city traffic.

The change from one power source to the other is automatic and seamless. The driver may be aware of the change, but does not have to initiate that change or even approve it.

Most hybrid cars will switch themselves off whilst the car comes to a halt and will fire up again whilst the accelerator is depressed. This one feature alone saves a lot of fuel. In traffic, the car is almost certainly using its battery-powered electric motor anyway, so it is very easy to stop and start it.

A hybrid car can be plugged into the national electricity grid to recharge its batteries, which may be necessary sometimes if the car is locked in traffic for a substantial part of the week. However, if you drive on long runs and in the city, that is give your hybrid car a balanced usage, the car will keep the batteries recharged by itself - mostly by the use of alternators and the braking system.

The aspirations of governments, environmentalists and drivers are being pinned on the more extensive use of hybrid cars and here are a few reasons why:

1) if the fuel efficiency of US cars was raised by one mile per gallon, it would save the total oil produced in the Arctic National Wildlife Refuge for two years.

2) the US would not have to import any oil at all from Kuwait or Iraq, if engine effectiveness was raised by 2.7 mpg

3) if US car fuel effectiveness was raised by 7.6%, then they would not have to import any oil from the Gulf at all.

Hybrid cars generally save over 7.6% on oil consumption, so the proliferation of hybrid cars and hybrid trucks could resolve the fuel and environmental crises being experienced by Western countries and eliminate our dependence on Arab oil.

Hybrid Cars

Hybrid Cars and Hybrid Trucks

The fuels that power the majority of cars and trucks, and indeed motorcycles and aeroplanes, is one of the most volatile commodities on the market. Oil and petrol are not just becoming more costly, but the crises in the Middle East threaten to disrupt supplies too.

This state of affairs is very disturbing for individuals and governments alike. As the cost of oil rises, citizens complain and blame the government and the rising cost of oil affects the cost of living and the balance of payments.

On top of that, environmental groups are undivided in blaming the consumption of fossil fuels, which includes oil, for most of the degradation of the environment and the consequential vanishing of species. A possible solution to all these woes is the development of a different kind of engine that does not burn so much oil. Enter the hybrid engine.

All the top car makers are busy making energy-efficient hybrid cars. Ford, Honda and Toyota are at the forefront of producing stylish cars that contain hybrid engines which run off petrol and electricity. In fact the car basically has two engines which share the mechanism for delivering the power to the wheels.

These cars use petrol while the batteries need charging or whilst the car needs extra power, say for overtaking or going up hill, but they automatically switch the petrol engine off and the electric engine on when electricity can provide enough power to achieve what you need the car to do, like cruising in city traffic or standard, unhurried driving. The batteries are charged by the petrol engine, by braking and by plugging it into the national grid.

Trucks obviously use a great deal more gasoline than cars and so the possibility for saving is a great deal higher to. The problem is that the electric motors are not really powerful enough to be able to completely take over from a petrol engine if a lot of power is needed to drive a fully laden truck.

It can 'assist' - that is reduce the load on the petrol

engine, thus saving some of the costs, but can it save enough fuel to justify it's fairly high price? That is the big question for all truck owners. However, the technology is being improved quickly and it almost certainly will do one day.

Again the big three are doing everything they can to compete in this potentially extremely lucrative market. If they could make hybrid engines that are powerful enough to pull a fully loaded truck at a decent speed, manufacturers are convinced that truck owners will go for them in order to save on their costly fuel bills.

This along with decreasing the cost of hybrid cars is the key to decreasing a country's dependence on imported oil. If you are not very worried about the high price of buying a hybrid car, then you should get one only to do your bit for the environment, but if you want to buy one to save on your fuel bills, you will have to get the calculator out and do your sums carefully.

Hybrid Cars

Hydrogen Fuel Cars - Do They Exist?

There are hydrogen fuel cars on the roads of some cities. However there are two ways in which hydrogen can be used to power cars. The first method is to use hydrogen to actually power the internal combustion engine, in much the same way as numerous cars use Liquid Petroleum Gas (LPG). The second way is to use the reaction of hydrogen with oxygen in fuel cells as a battery, which makes the car a kind of electric car.

The dream of creating hydrogen in the car while driving along by electrolyzing water is still some way off, so we are still at the stage of batteries and filling the tank with hydrogen gas. This is the nub of the problem for potential users and manufacturers. There are just sixteen hydrogen filling stations in Los Angeles and none in 99% of other cities worldwide.

In fact, some of the big name automobile manufacturers have pulled out of the race to put the first commercially viable hydrogen powered car on the streets. Ford and GM have declared that they are pulling out in America and so has Renault in France.

However, the Japanese firms are pressing on. In fact, Honda introduced its first hydrogen fuel cell car in 1999. It was known as the FCX and they are now ready with introductory models of the second generation hydrogen cars known as the FCX Clarity. Guess where they are available for sale? The one city in the world? Yes, Los Angeles, because of its hydrogen stations.

Honda says that, they could go into full-scale production of the FCX Clarity by 2020, if the world is prepared for them by then. Hyundai have on-going plans to manufacture fuel cell (FC) cars and say that they will be in place to launch full-scale production by 2012. Daimler also declared that they would be manufacturing 100,000 FC vehicles in 2012-2013.

Then there are hydrogen powered buses in quite a few European cities including Amsterdam, Barcelona, Hamburg, London, Luxembourg, Madrid. Porto Stockholm and a few more. Lotus, the makers of London taxis, have announced that they intend to set up a fleet of new, hydrogen powered taxis in time for the London Olympics in 2012.

Consequently, the hydrogen vehicle and the hydrogen passenger car is out there and the numbers will be growing fairly soon. The buses, talked about above, go back to their depot, where an electrolyzing machine converts water into fuel for them to fill up on and the same will be the case for many of London's taxis. Regrettably, procuring fuel is not the only difficulty for the average motorist, a number of these vehicles, like the FCX Clarity cost about $300,000 each.

However, here are a few interesting facts for those who enjoy trivia. Francois Isaac de Rivaz designed the first hydrogen-powered car in 1807 and Paul Dieges filed a US patent for a modification to the internal combustion engine in 1970, which enabled a petrol engine to run on hydrogen and 200 years later we are still trying to get it right.

Hybrid Cars

New Hybrid Vehicles

New hybrid vehicles are environmentally friendly because they emit 90% less toxic gases, which also makes them beneficial to our environment and our health.

Scientists believe that new hybrid vehicles do not contribute much to air pollution so the use of hybrid vehicles helps reduce the rate of global warming. Hybrid vehicles also have reduced oil dependency, which increases fuel efficiency by 15% to 30%.

If the price of new hybrid vehicles were to come down, then, I think that most people would consider buying one. However, the reality is that they are still extremely expensive relative to other cars.

Having said that, it is also true that hybrid vehicles hold their value more than conventional cars and the savings in fuel make a new hybrid vehicle a viable proposition.

If, and it is a huge if for the vast majority of people, if you do not need finance for that new hybrid vehicle! If you do, then the benefits of having that hybrid are trickling through your fingers to the bank or the finance company, which is a real shame.

It's a real shame for the environment. Not for the hybrid vehicle manufacturer, not for the finance company! They are still doing alright, thank you very much.

So, it comes down to government sponsorship of the hybrid vehicle manufacturers. And, to be fair, some governments, like the government of the United States of America is trying to help, but it is treading carefully. Too carefully.

(Look for an article on this site that examines this aspect of government policy a little more deeply).

New hybrid vehicle owners can save tax. The IRS has offered large tax credits for anyone purchasing a new hybrid vehicle. But it is important to remember that these are tax credits not tax reductions.

The amount of tax to be paid depends on the model of the particular new hybrid vehicle being purchased. The law states that every car manufacturer is only allowed to sell 60,000 new hybrid vehicles at the new credit limit, which means that top hybrid models could easily be sold out before the date the system expires ie Dec 31, 2010.

So, why should you buy a new hybrid vehicle? Well, new hybrid vehicles and other forms of new hybrid transport are as reliable and comfortable as conventional vehicles; parts are not as expensive as they once were; there are tax credits and fuel economies and hybrids are environmentally friendly.

With all these reasons for getting a new hybrid vehicle, there is no real excuse for writing hybrids off out of hand. We have many articles and resources about new hybrid vehicles on this website and hope that they will help you make up your mind

Hybrid Cars

Hybrid Electric Engines

Have you ever wondered about hybrid cars? They certainly look stylish and modern, but what is it about them that makes them so different and so special? After all, many cars look the same these days, do they not?

They are all designed by computer to be aerodynamic and aerodynamic is aerodynamic, so they all end up looking indistinguishable to anyone who is not an enthusiast.

Well, the fact is that the difference between a hybrid car and a normal car lies under the bonnet. There are different kinds of hybrid vehicles as well, but the most common sort has a hybrid electric engine.

Therefore, you could say that what makes hybrid cars different from most other, ordinary cars is their engine.

Hybrid Cars

Or engines really, because hybrid electric cars basically have two engines. One, the internal combustion engine, runs on petrol, LPG or diesel and the other, the electric motor, runs off electricity stored in an array of powerful batteries.

On the other hand, normal cars only have an internal combustion engine. it is a good idea to examine these two types of engine separately.

First of all the electric motor. Electrical energy is created by various components on the car and stored in a large bank or array of high-capacity batteries. The electricity is created by the turning of the internal combustion engine, just as with a normal car, when it is in use, but also by the wheels or / and the car's braking system.

Different models have different ways of generating this electricity, but they are all very hi-tech and very efficient.

When the car's internal computers decide that the car no longer needs the power of the internal combustion engine, it will turn it off and switch to the electric motor instead.

The circumstances where this might happen are, for example, when you are driving slowly in inner-city traffic, when you are cruising at moderate speed on a flat motorway or when you are going down hill. This saves a considerable amount of fuel, which is obviously a great saving to you.

However, there are times when you need more power that the batteries can give you, for example, when accelerating hard, when overtaking or when climbing a hill.

At times such as these, the electric motor will shut down and the combustion engine will take over and start supplying power to the accelerator and electricity to the batteries. When the need for extra power has passed, the batteries will take over again.

These decisions are all taken by the hybrid electric vehicle's (HEV's) on board computer system and you will not notice much other that a surge of power of a quietening of the propulsion system.

This works very well in most scenarios, but some drivers of HEV's would like to have a manual override for unusual circumstances, such as an undulating terrain.

All in all, cars with hybrid electric engines work

very well at reducing fuel bills but they are still too expensive to be much more than a gimmicky toy and a salve to the conscience for the rich.

The Hybrid Car and Petrol Prices

The appeal and popularity of the hybrid car have grown exponentially, especially with the rising worries about high petrol prices together with worsening air pollution.

Here are some interesting and, I hope, useful bits of information that might assist you to learn more about the hybrid car and how they may help you save money on fuel and be somewhat protected from rising fuel costs.

A hybrid car is the type of car, or any other vehicle, that makes use of at least two different fuel sources to make it run. Both fuel sources are sometimes used in unison to help propel the vehicle more efficiently.

There are several different combinations of hybrid car possible, but the most common hybrid car so far is the petrol-electric hybrid.

The petrol-electric hybrid car, also known as the hybrid electric vehicle or HEV, makes use of a gasoline internal combustion engine or ICE and a separate electric motor to power it. While the ICE makes use of gasoline to make it go, an electric battery is used to store the electrical energy that powers the hybrid car's electric motor.

The HEV usually has a gas engine that is smaller in size and weight than the conventional one used in standard petrol powered cars. Use of more advanced technology makes this possible and allows the HEV to have better running efficiency as well as substantially reduced polluting emissions.

Apart from the gas engine, the hybrid electric car also has a specially designed electric motor built in that not only provides additional power to the car but also acts as a generator when it is not being used in drive mode.

The electric motor acts as a generator, in situations when it is not being used to drive the hybrid car, to help charge the battery for added efficiency. Even the braking system is used to charge the battery.

In the usual HEV set up, the car uses its electric motor when being driven at very low speeds, say, in traffic jams. The petrol engine acts as a secondary power source when the HEV requires much more power, such as when climbing a hill.

The petrol engine also supplies the electric motor with power whenever the car needs it in order to go faster, such as when overtaking.

The petrol and the electric motor can also work together in certain circumstances, when needed. Because the hybrid electric car makes use of both an electric motor and a petrol motor, a substantial improvement in car mileage is achieved.

A hybrid electric vehicle or HEV can go longer distances using the same amount of fuel compared to a traditional petrol powered vehicle. Whenever the electric motor is being used, petrol consumption is reduced.

This results in less petrol being used when travelling the same distance as a traditional, petrol powered vehicle.

Furthermore, because the hybrid electric car has a smaller petrol engine, the hybrid car also runs more efficiently because of less engine weight compared to a conventional car's engine. The working parts of the hybrid car engine are also smaller and require less energy to move them.

This efficiency makes the hybrid electric car quite a good option for people concerned about rising gas prices. Using a hybrid car can help drivers save a substantial amount of money when travelling.

Not only that, using the hybrid car can also help to reduce polluting emissions by using less petrol while on the move, especially in heavy traffic.

Electric Hybrid Cars

There are different modes of transport that people can use. Among those is the electric hybrid car. The various makes of electric hybrid car allow people the choice of having a stylish looking vehicle, which at the same time conserves petrol.

Additionally, the environment is kept more pollution free by these electric hybrid cars. The types of hybrid cars which you see included in the range of electric hybrid cars will be the popular SUVs, sports cars and pickup trucks.

You will have the chance to see those latest electric hybrid cars, which are due to come on to the market. As there are many types of electric hybrid cars you should look to see if there are any performance reviews available.

This knowledge will help guide you when you are trying to make up your mind about which electric hybrid car to buy. You will need to realise that some of the electric hybrid cars will not be that easy to repair or have their components replaced. This is one factor that you should keep in mind while you are shopping for the electric hybrid car that suits you.

Sure, there will be some times when you will not be able to uncover the exact model of electric hybrid car that you require. In order to reduce this frustration, make a list of several different electric hybrid car types. Try including some of the better known electric hybrid car models).

You may also want to make a list of the most important qualities that you are looking for from an electric hybrid car. Don't forget that you ought to visit the car showrooms to inspect the electric hybrid car of your choice, since I'm sure you may appreciate seeing what sort of hybrid vehicles are available before you buy one.

Because there will be occasions when you can not get hold of the details of all of the electric hybrid cars on the market, you should find some other way of getting this information.

Out of the many means that you can use to look up the many models of electric hybrid car, the Internet can provide you with pictures, specifications, descriptions and reviews of the different makes of electric hybrid car.

For instance, you can obtain the information for the currently available batch of hybrid Honda cars or the previous models too.

Amongst the plethora of information that you can look up are: the number of seats and the safety and engine specifications of these hybrid cars. There are, as we all know, many different types of hybrid car that can be bought.

These will, for the most part, be well-known car makes. You will find that Honda which is a world-famous manufacturer of cars with modern technology also has their version of a hybrid car. The electric hybrid car is one of their hybrid cars that is very popular with the public.

Hybrid Cars

Run Your Car on Water

Every now and again, stories surface in the media saying that we can convert our cars to run on water and gas to save over 40% on our on fuel costs. Have you seen them? The problem is that we never get told how to do these conversions, or the name of the companies that can carry them out for us.

However, by all accounts, it is possible to convert your car to a water-burning car at least in theory. Many people say that they are working on ways for us to be able to run our cars solely on water, or as a supplement to another power source such as electricity or petrol, which they say would increase our cars' fuel efficiency and reduce your fuel costs significantly.

Some add that they face a great deal of opposition from the government-sponsored grants councils and the oil companies, who are worried about the reduced profits that lower sales would bring. Some people say the oil companies buy up and mothball any invention that looks likely to become commercially viable.

I for one can believe this. Think about a 40% drop in revenue for the oil companies and a 40% reduction in tax revenue for the government. It would devastate them, no matter what the government says about it wanting to save the planet and reduce imports.

The simple and easy explanation, of the theory for burning water says to use water from a reservoir in your car and pass electricity from your car's battery or motor through it to separate the water into a
gas called HHO (2 Hydrogen + 1 Oxygen). HHO, also called Brown's Gas or Hydroxy, burns smoothly and provides
significant energy - while the end product is just H_2O! No smelly, poisonous carbon monoxide exhaust - just a few dribbles of water which could be redirected back into the car's reservoir!

HHO provides the 'atomic' power of Hydrogen, while maintaining the stability of water.

So, the question is: can Water really power a car? And the answer is 'YES, most definitely!'.

Apparently the technology to build a water-burning hybrid engine is easy and affordable.

Water can be used to fuel a car when used as a supplement to a petrol generator, since using only batteries to provide all the electricity s proving difficult. The adversely affect the power-weight ratio of the car too much, as other hybrids have found. Still great minds are working on the problem. see Elon Musk's Tesla project.

In fact, very little water is needed! On report states that just one quart of water can provide over 1800 gallons of HHO gas which can literally last for months and significantly increase your vehicle's fuel efficiently, improve emissions quality, and save you money, and upset the oil companies.

I cannot tell you more here, because the information would soon be out of date. However, if this fascinating subject interests you as well, search for 'hydrogen engines' on a search engine and YouTube and subscribe to Musks's Tesla newsletter.

Some industry insiders say its just a matter of time before this water-burning technology will be standard in new automobiles. One expert estimates most cars will be using this technology by 2025.

Hybrid Car Myths

If you are thinking about a hybrid car, you may be hearing quite a bit of "talk". Some people think the hybrid car is the best thing on the market. Some people say it's just a phase, because of falling oil prices. Other people say they think you can save a lot of money, but you're not sure whether it's really worth it.   What's the truth, and how do you separate myth from fact with all of the stuff that is being thrown at you? Below, you can read and understand the common hybrid car myths.

<u>Hybrid cars are the same as electric cars</u>

This is not true because hybrid cars are fuel-powered for the most part. They have what are called battery assists. The assist is powered by a nickel-metal hydride battery pack that is rechargeable. t takes over under easy conditions.

<u>You are guaranteed to save money with a hybrid car</u>

If you are doing mostly city driving, you may save petrol and you may not. The same goes for highway driving. There are just many different factors such as the terrain of the city, since electric power is not much good on a steep hill. It has been said that if everyone bought a hybrid car, petrol consumption would decrease by only 10%. That's not a very big difference.

A hybrid cars battery can run out

A hybrid car's battery should not run out while you are driving it. The engine in a hybrid car does not idle when stopped (at a red light for instance.) What does it do instead? It recharges its battery. So there's no need to worry about a hybrid car stopping you.

The hybrid cars rechargeable battery only lasts for two years

A hybrid car certainly would not be worth purchasing if this was the case. A hybrid car's rechargeable batteries usually come with an eight to ten year warranty.

If I run out of petrol, I can keep driving on the hybrid car's battery

Keep in mind, a hybrid car's battery is an assist. That means that hybrid cars still run on fuel. After you run out of gas, the battery may keep the car running for just a little while. However, the car will stop very soon.

Hybrid cars will soon put conventional car sellers out of business

This probably won't happen anytime soon. The reason for the delay has to do with the cost of a hybrid car. Many people simply can't afford one. Also, people just aren't too sure whether they will really save money on hybrid cars. Therefore, they are slow to join the people who want a hybrid car. As for business, traditional car manufacturers are the main source of hybrids.

Hybrid cars will only save you about ninety dollars a year

I did hear something on the news about this once, but it may not be true. If there's something you really want to know though, and there's a lot of smoke surrounding it, you simply have to start digging and do some of your own research. There are many different models of hybrid cars, and many different manufacturers make them. This means that there may be many more variables involved than the ones discussed here. A hybrid car may help you, and it may not, but the final decision is up to you.

Hybrid Cars and the Energy Crisis

It used to be said that not enough people were doing all they can to fight against the energy crisis. That crisis has all but disappeared in some countries now, but hybrid cars can still help further. However, their high cost means that not enough people are driving them. Here are a few issues related to the then energy crisis and how hybrid cars can help.

<u>The West isn't doing enough</u>

The United States consumes the most fossil fuels in the world. Most people believed that all of our energy problems can be solved if we would only look further into the oil deposits in Alaska or if we made full use of the recent oil discovery in the Gulf of Mexico. In the end, it was shale gas and oil that came to the rescue.

Countries still import vast amounts of oil, so possibly hybrid cars could make it so that we might not have to use other sources of energy at all to keep the economy going, because hybrid cars cause drivers to use even less fossil fuel.

Energy consumers just accept petrol prices

People used to care that petrol prices were much higher than they were years ago. Now, after all the oil produced from fracking, the price has not returned to what it used to be. People just accept the high prices. In the meantime, cars are getting bigger and bigger.

Car manufacturers are making trucks and SUVs. These cars use more petrol, but you won't believe how many people just won't give up their dear old SUV. Hybrid cars end up costing people less to own than conventional cars do. So there's no need to worry about just settling for being swindled by the oil economy.

Soon there could be an end to the "cheap oil period"

Soon, we could all be in over our heads, but not because of an energy crisis, but a climate change problem. That will mean that everyone, not only drivers will have a problem. However, despite the satisfaction of drivers with the low oil prices, one day, countries may still battle over who gets oil and who doesn't. Still, for now, the peak oil crisis can be put off, if more people would just purchase hybrid cars. The good news is that advances are being made to hybrid cars everyday.

<u>The plug-in hybrid car for instance, might one day, not need oil at all</u>

So in the event that we have another energy crisis, people should really use the time to get their heads together and create a unified fight against conflict over oil. Hybrid cars are one way to delay the energy crisis until scientists can come up with another alternative, like water-powered cars.

So that's it, the Western mass use of the hybrid car might make it so that th world needs less oil. One problem that remains though, is that the hybrid manufacturers are still charging too much for their hybrid cars.

Hybrid Cars

Hybrid Cars vs Plug-in Hybrid Cars

Hybrid cars are on everyone's minds. The high cost for a full tank of petrol caused many people to complain loudly, until the governments had to prove that they were listening and doing something about it, if they wanted to get elected. As hybrid cars started to roll off the production lines, people applauded for the small amount of gas they need to operate them, but sighed at the high cost of the hybrid vehicles themselves as they drove them off the lots of car dealerships each and everyday.

They are still expensive, especially given the drastic fall in the cost of oil, but what about a plug-in hybrid? Most consumers have heard that these cars are great too. Then, a person might ask him or herself, what exactly is a plug-in hybrid? How do they work, and what's the difference between a plug-in hybrid and a regular hybrid?

Plug-in hybrids are able to run solely on batteries,

but they use fuel also

This type of hybrid car has some of the characteristics of petrol hybrid vehicles. They are also very similar to electric vehicles. Plug-in hybrid cars must be recharged externally by connecting a plug to a power source. The combustion engine in plug-in hybrid vehicles is used only as a back up. These cars can run only on batteries if desired. However, it is expected that these types of hybrid cars be plugged in daily or, more often, overnight.

Hybrid cars travel just as many miles as a conventional car

Designed to go the extra mile where gas-mileage is concerned, hybrids can be driven on the highway, in cities, or wherever else a person needs to drive. On the other hand, plug-in hybrids are designed to handle commuter-type distances, meaning about twenty to sixty miles between recharges. This way, the plug-in hybrid does not have to use its back up combustion engine, but plug-in hybrids can go further using petrol as well.

Hybrids help to minimise pollution, but they still pollute the air

When compared to plug-in hybrids, hybrid cars

have a long ways to go where pollution is concerned. Since plug-in hybrid cars can run solely on their battery power, they don't have to emit poisonous waste all the time. That means that plug-in hybrids don't have to pollute the air as much.

Plug-in hybrids fight against greenhouse gases

Plug-in hybrids use virtually no petrol. Board studies have shown that electric hybrids emit at least 67% less greenhouse gases when compared to gasoline cars. Since the products used to power plug-in hybrids are renewable, the difference in greenhouse gases may be even greater than the study determined. However, if fossil fuels are used to produce the electricity, then the savings in exhaust are reduced.

And there you have it. That's the difference between plug-in hybrids and regular hybrid cars. It makes a big difference, but you would be surprised how little that matters at the current moment. And that's only because plug-in hybrids are not being sold to consumers in great numbers at this time. However, this list should get you excited about the wonderful plug-in hybrid car.

People already really like regular hybrid cars, but

they haven't seen anything until they test drive plug-in hybrid cars. However, the costs are still high, the cash saving lower than ever, and who knows? Hydrogen cars might be just around the corner.

Owen Jones

The Future Landscape of Hybrid Cars

In the ever-evolving realm of automotive innovation, hybrid cars have carved a significant niche, offering a tantalizing glimpse into the future of sustainable mobility. As a car enthusiast, the prospect of what lies ahead for hybrid vehicles is nothing short of exhilarating, with advancements in technology and shifting consumer preferences steering the automotive industry toward a greener horizon.

1. Advancements in Battery Technology:

The heartbeat of any hybrid car is its battery, and the future promises remarkable strides in this critical component. As research and development accelerate, we anticipate batteries with higher energy density, longer ranges, and faster charging times. The evolution from traditional lithium-ion batteries to solid-state batteries holds particular promise, potentially providing increased efficiency and safety while reducing the environmental impact of manufacturing and disposal.

2. Integration of Artificial Intelligence:

The likely future of hybrid cars is intertwined with the seamless integration of artificial intelligence (AI). Smart algorithms will optimise energy usage, predicting driving patterns and adapting the powertrain accordingly. AI-driven systems will not only enhance fuel efficiency but also contribute to a more personalised driving experience. Imagine a car that learns your preferences, adjusts performance settings, and even suggests the most energy-efficient routes—all in real-time.

3. Hybridization Across Vehicle Segments:

While hybrid technology initially found its place in compact cars and sedans, the future heralds its expansion across diverse vehicle segments. From SUVs to high-performance sports cars, hybridization is becoming a universal language in the automotive lexicon. The marriage of electric and combustion power is not just about fuel efficiency; it's about delivering a dynamic driving experience that caters to a broad spectrum of automotive enthusiasts.

4. Hybrid as a Stepping Stone to Full Electrification:

Hybrid cars are poised to act as a pivotal bridge in the transition toward full electrification. As charging infrastructure continues to develop, hybrids offer a practical compromise, alleviating range anxiety and catering to those who may not be ready to fully embrace electric vehicles. The likely future sees hybrids as a stepping stone, providing a familiar and accessible path for consumers to join the electric revolution gradually.

5. Eco-Friendly Materials and Sustainable Design:

As environmental consciousness takes center stage, the future of hybrid cars extends beyond their powertrain. Manufacturers are exploring eco-friendly materials and sustainable design practices to reduce the overall environmental footprint of vehicles. From recycled interior materials to innovative manufacturing processes, the next generation of hybrids aims to embody a holistic commitment to sustainability.

6. Performance-Focused Hybrid Models:

Contrary to the misconception that hybrids sacrifice performance for efficiency, the future promises an array of performance-focused hybrid models. Hybrid technology, with its instant torque delivery, can enhance acceleration and handling, making these cars not just eco-friendly but thrilling to drive. The racetrack may soon see hybrids competing at the highest levels, challenging the preconceived notions of what a high-performance vehicle can be.

7. Government Incentives and Policy Support:

The likely future of hybrid cars is intricately linked to government incentives and policy support. Many countries are introducing stringent emissions regulations and promoting eco-friendly technologies through financial incentives. This support encourages both manufacturers and consumers to embrace hybrid solutions, fostering a more sustainable automotive landscape.

8. Affordability and Mass Adoption:

As technology matures and economies of scale come into play, the cost of hybrid cars is expected to decrease. This trend, coupled with the growing awareness of environmental issues, will likely drive mass adoption. Hybrid vehicles will not remain confined to a niche market but will become an accessible and practical choice for a broader spectrum of consumers.

In conclusion, the future of hybrid cars is a thrilling trajectory that combines technological innovation, environmental stewardship, and a commitment to delivering an unparalleled driving experience. As a car enthusiast, the evolving landscape of hybrids promises not just a greener tomorrow but a dynamic and exciting journey toward automotive sustainability. The road ahead is paved with innovation, and hybrid cars are set to lead the way into a new era of automotive excellence.

Hybrid Cars

Contact Details

Facebook: AngunJones
Twitter: @owen_author
Blog: Megan Publishing Services

This book is part of the 'How to...' series of 150 manuals by Owen Jones.
The whole series can be found in many languages on Megan Publishing Services at:
https://meganthemisconception.com

www.ingramcontent.com/pod-product-compliance
Ingram Content Group UK Ltd.
Pitfield, Milton Keynes, MK11 3LW, UK
UKHW021645190726
13853UKWH00001B/63